AF443533

Titles in Series S8711

My book of nursery rhymes

My book of animal rhymes

My book of playtime rhymes

My book of bedtime rhymes

My alphabet book

My counting book

My book of colours and shapes

My book of opposites

British Library Cataloguing in Publication Data
Bradbury, Lynne
    [Colors and shapes]. My book of colours and shapes.
    1. Color—Juvenile literature
    2. Geometry—Juvenile literature
    I. [Colors and shapes]   II. Title
    III. Burton, Terry
    516   QC495.5
    ISBN 0-7214-9569-9

First edition
Published by Ladybird Books Ltd Loughborough Leicestershire UK
Ladybird Books Inc Lewiston Maine 04240 USA

© LADYBIRD BOOKS LTD MCMLXXXVIII

All rights reserved. No part of this publication may be reproduced, stored in a retrieval system, or transmitted in any form or by any means, electronic, mechanical, photo-copying, recording or otherwise, without the prior consent of the copyright owner.

Printed in England

My book of
colours and shapes

by LYNNE BRADBURY
illustrated by TERRY BURTON

Ladybird Books

Micky and Max are painters.
They are going to paint this house.
What a mess it is!

Micky gets a long ladder.
Max gets the brushes.

Now for the paint.
First they get a tin
of red.

Next they get
some blue,
and then
yellow paint.

''We'll paint each shape a
different colour,'' says Micky.

"What's a shape?"
asks Max.

''All these round windows
are circle shapes,'' says Micky.
''Paint them red, please.''

Max paints the circles red.
But he misses one. Can you see it?

''Now,'' says Micky,
''do you see those windows,
with four sides all the same?
They're squares. Paint them blue.''

Max paints the
square windows blue.
But he misses one!

"What about the doors?" asks Max.
"They have two long sides
and two shorter ones."

"Those are rectangles," says Micky.
"Paint them yellow."
Max does. But he misses one!

"We've run out of colours," says Max.
"No, we haven't," says Micky.
"Watch this!"

Micky mixes red and yellow.
''This colour is called orange,''
he says.

Micky points to the roof.
''Those shapes with three sides
are triangles,'' he says.

''Paint them orange.''
Max does. But he misses one!

Next Micky mixes blue and yellow.

''This colour is called green,'' he says.

Max paints the roof green.

''Mixing colours looks like fun,'' says Max.
''Can I try now?''
Max mixes red and blue paint.
''You've made a colour called purple!''
says Micky.

red
purple
blue

Max paints the rest of the house purple. ''Finished!'' he says.

''Oh, no!'' says Micky.

Can you see all the shapes Max has missed?
What colours should he have
painted them?

Look back in the book to find out!